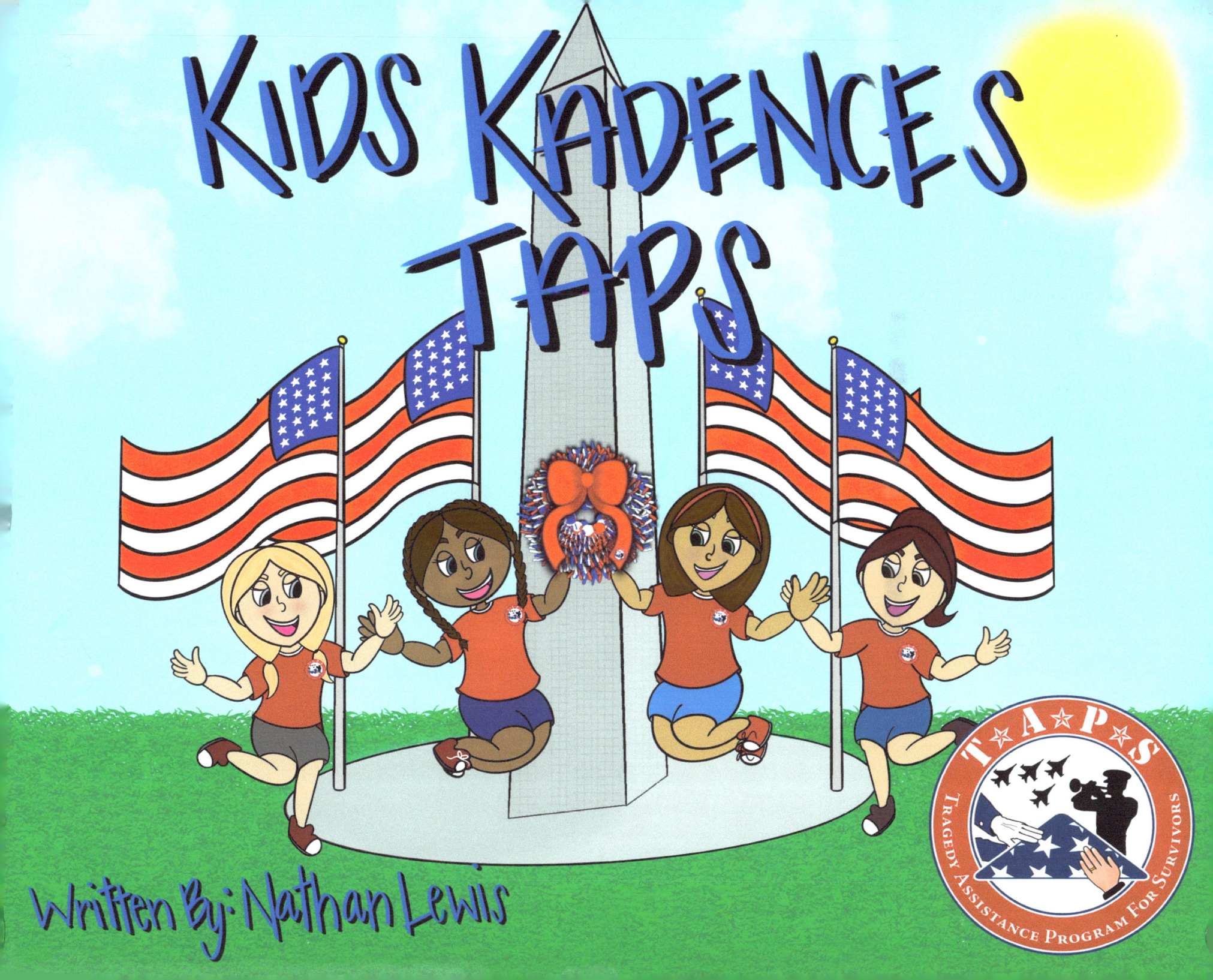

Kids Kadences Taps

Written By: Nathan Lewis

TAPS — Tragedy Assistance Program For Survivors

Copyright © 2022 Nathan Lewis and Sky Soldier Publishing

All rights reserved. No part of this publication may be reproduced, distributed, or transmitted in any form or means, including photocopying, recording, or other electronic or mechanical methods, without the prior written permission of the publisher or TAPS, except in the case of brief quotations embodied in critical reviews and certain other noncommercial uses permitted by copyright law. For permission requests, email the publisher at kidskadences@gmail.com.

ISBN: 979-8-9855367-0-6 (Hardcover)
ISBN: 979-8-9855367-1-3 (Paperback)

Front cover image and art work by Jansen Morgan.
Book design by Jansen Morgan and supported by Blue Pen Books.

Published by Sky Soldier Publishing LC

TAPS Copyright: TAPS is a National Nonprofit 501(c)3 Veterans Services Organization and is not a part of, or endorsed by, the Department of Defense.

Tragedy Assistance Program for Survivors Mission: TAPS provides comfort, care, and resources to all of those grieving the death of a military loved one. Since 1994, TAPS has provided comfort and hope 24/7 through a national peer support network and connection to grief resources, all at no cost to surviving families and loved ones. TAPS provides a variety of programs to survivors nationally and worldwide. Our National Military Survivor Seminar and Good Grief Camp has been held annually in Washington D.C., over Memorial Day Weekend since 1994. TAPS also conducts Regional Youth Seminar for Adults and Youth Programs at locations across the country, as well as, retreats and expeditions around the world. Staff can get you connected to counseling in your community and help navigate benefits and resources.

For more information please visit www.taps.org

100% of all proceeds from the purchase of this book go to TAPS.

This book can be read as is with a normal rhyming rhythm, but it is best when read in a rhythm and tone similar to the military cadence C-130.

Please use your smartphone to scan this code and watch the video on YouTube of Army Soldiers singing the Cadence for you to hear the rhythm it was intended for or visit https://www.youtube.com/watch?v=HL4yToiP6VE

TAPS airplane rolling down the strip

64 TAPS kids on a little trip

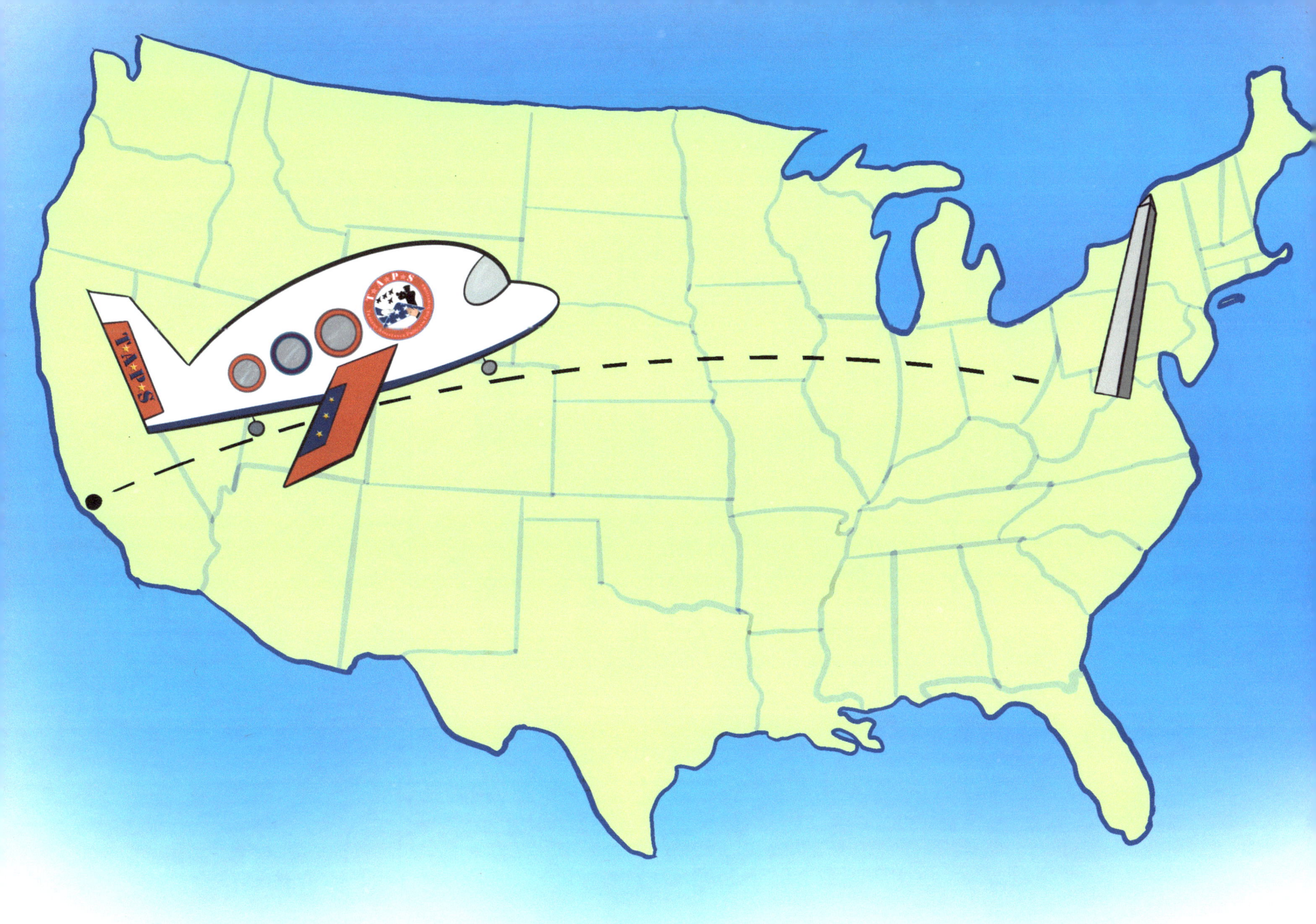

Mission is vital, destination is known

Good Grief Camp, we don't want to go home

Pack up, load up, don't forget your pin

TAPS shirt, shorts, and your swim fins

Get to the car by the count of four

My knees feel the breeze as I go out the door

If the driver forgets the way

GPS will save the day

If our vehicle breaks down

Our mentors will bring us to town

Good grief camp we're here at last

It is time to remember our past

All the TAPS children come together to play

In Loving Memory of

Honoring our Heroes is the only way

From the Author

I am honored to be a part of the TAPS family for the last decade as a Mentor, Volunteer, and Group Leader for these amazingly resilient young people.

Thank you to all of the incredible Mentors, Group Leaders, Support Staff, and Volunteers that make these amazing TAPS events possible.

Thank you to my wife, Brooke, who supported me in this 25 year Army journey from starting as an Enlisted Logistician in the Army Reserves through the University of Toledo and into the Active Duty in Army Aviation and Military Intelligence, onto Georgetown University and then academia at Xavier University and Loyola University Chicago.

My most sincere appreciation is for Bonnie Carroll and all the Gold Star Families whose strength and resilience have redefined our understanding of service and sacrifice.

From the Illustrator

I am honored to have the privilege to illustrate this book for this amazing organization. TAPS does a lot of amazing work for people all around the world and to be able to illustrate this book for them truly is a huge honor.

Thank you to Nathan and his wife Brooke for always supporting me in my life and giving me this opportunity. Thank you to my family for always giving me the tools to pursue my love of art in the many forms I do. Thank you to my amazing friends for always encouraging me and being my support group in both the good and the bad.

To anyone who has purchased this book thank you for supporting this beautiful organization.

CPSIA information can be obtained
at www.ICGtesting.com
Printed in the USA
BVHW010052030523
663457BV00001B/1